The Chemistry of THE RADIO ELEMENTS

PART 2

MAXWELL PRESS

The Chemistry of THE RADIO ELEMENTS

PART 2

Frederick Soddy, F.R.S.

Winner of Nobel Prize for Chemistry
Lecturer in Physical chemistry and Radioactivity in the
University of Glasgow

MAXWELL PRESS

Chennai Trichy NewDelhi

MAXWELL PRESS

ISBN 978-93-90063-46-8 Maxwell Press

Printed and bound in India
No. 44, Nallathambi Street,
Triplicane, Chennai 600 005

MJP 892

Publisher : C. Janarthanan

Publisher's Note

The legacy of a country is in its varied cultural heritage, historical literature, developments in the field of economy and science. The top nations in the world are competing in the field of science, economy and literature. This vast legacy has to be conserved and documented so that it can be bestowed to the future generation. The knowledge of this legacy is slowly getting perished in the present generation due to lack of documentation.

Keeping this in mind, the concern with retrospective acquiring of rare books has been accented recently by the burgeoning reprint industry. Maxwell Press is gratified to retrieve the rare collections with a view to bring back those books that were landmarks in their time.

In this effort, a series of rare books would be republished under the banner, "Maxwell Press". The books in the reprint series have been carefully selected for their contemporary usefulness as well as their historical importance within the intellectual. We reconstruct the book with slight enhancements made for better presentation, without affecting the contents of the original edition.

Most of the works selected for republishing covers a huge range of subjects, from history to anthropology. We believe this reprint edition will be a service to the numerous researchers and practitioners active in this fascinating field. We allow readers to experience the wonder of peering into a scholarly work of the highest order and seminal significance.

MAXWELL PRESS

CONTENTS

THE CHEMISTRY OF THE RADIO-ELEMENTS

PART II

INTRODUCTION

"THE Chemistry of the Radio-Elements" dealt with the subject from a standpoint more or less original at the time it was published in 1911, and certain views were expressed tentatively (I, pp. 24–30) more as an individual opinion than a settled judgment, which have since developed. The extension not only constitutes a great simplification of the subject itself, but also has the most intimate and fruitful connection with the theories of chemistry in general, and in particular with the Periodic Law. The present monograph attempts to deal with these advances briefly in the form of a continuation or second part of the original one.

In the first place, the method of treating each radio-element as the chemical analogue of one or other of the known elements, from which it could not be separated and with which it was, as far as could be seen, chemically identical, whereby the chemical character of the radio-elements in question could be "accurately described in a single sentence," can now be shown to be general. Probably not one of the numerous new radio-elements possesses a unique chemical character unshared by others, and the chemistry of the thirty-four individuals now recognised becomes in consequence the chemistry of a much smaller number—about ten in all—of types of elements. Of these, five were known long before radioactivity, and the other five have been subsequently recognised as the direct result of that discovery.

Secondly, the generalisation (I, p. 29) that the expulsion of the α-particle caused the element to shift its position in the Periodic Table by two "places" in the direction of diminishing mass, so that the product is not in the next family, but in the next but one, has also been shown to be general, whilst the significant fact that the passage through the table appeared to be *alternating*, the matter passing often through the same family not once, but *twice*, has also been explained. In the β-ray changes it has been established that the element shifts its position in the Periodic Table by one place in the opposite direction to that for the α-ray change, so that a succession of three changes in any order of which two are accompanied by the expulsion of a β-particle and one by an α-particle must result in the product again coming back to the original place occupied by the parent. These laws are of perfectly general application, and hold, so far as can be seen, without exception. They establish a direct connection between the sequence of radioactive changes and the chemical character of the products, so that if the latter is known a valuable check is afforded as to whether the sequence of changes has been completely unravelled. Conversely, the sequence of changes can be used to determine completely the chemical character of the products, when for any reason, such as excessive instability, as in the case of the C′ members, or stability, as in the case of the end products, they remain experimentally unknown. It has already proved itself a guiding principle in clearing up doubtful points, and in greatly limiting the number of possibilities that must be taken into account in unexplored fields.

This important discovery is shared by several investigators, notably A. Fleck, A. S. Russell, and K. Fajans. [1, 2, 3, 4, 5.] It dates from the commencement of the present year, 1913; but new as it is, it has not been found advisable to deal with it at all in historical order of sequence. Though it must not be forgotten that all the necessary main advances in our knowledge of the chemistry of the radioactive elements, and of the sequence of radioactive changes, were made prior to the recognition of the underlying generalisation, and were not in any sense made to suit the needs of this generalisation (compare Annual Reports of the Progress of Chemistry,

Radioactivity, 1912). At the same time it has been possible to predict from the generalisation many new details, and these have been almost immediately and conclusively established by experiment.

In consequence we can now write the three disintegration

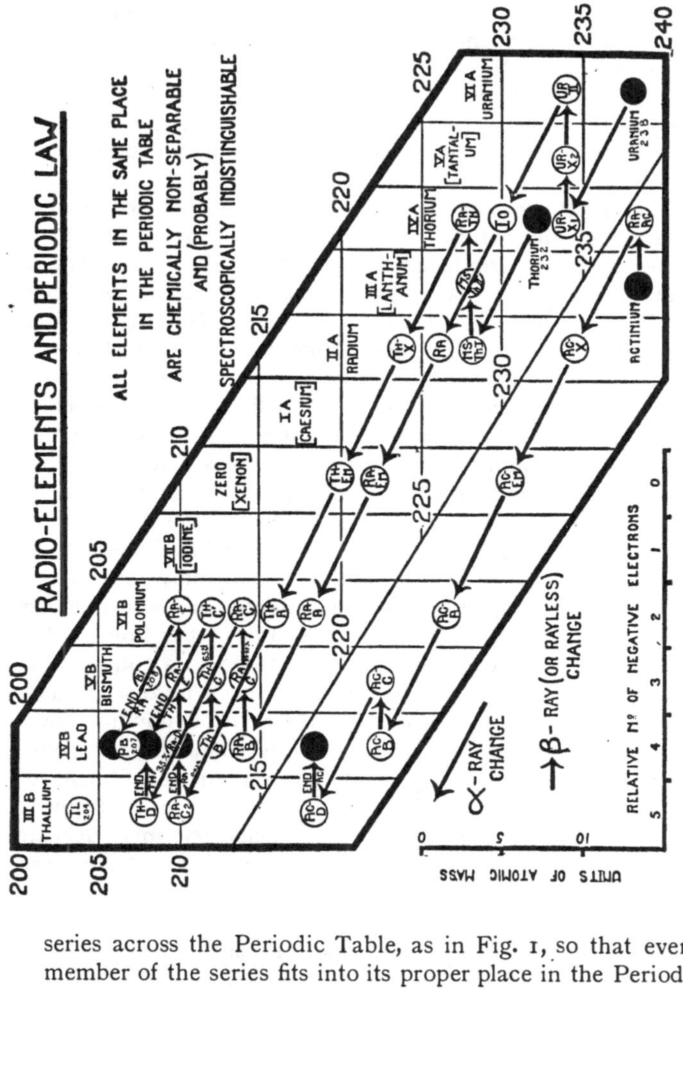

FIG. 1.

(*Reproduced by the courtesy of the British Association, from a Paper read in Section B. Birmingham, 1913.*)

series across the Periodic Table, as in Fig. 1, so that every member of the series fits into its proper place in the Periodic

Table ; and, so far as can be ascertained, all the elements occupying the same place are chemically non-separable and identical. The diagram is interesting, not only as establishing a fair presumption that the whole of these new changes have now been disentangled, but even more interesting because of the new light it sheds on what the Periodic Law means, and what the chemical elements themselves really are. The chemical non-separability and identity of different radio-elements, first definitely drawn attention to in the case of radium and mesothorium I,[6] and now shown to be the general characteristic of the radio-elements, becomes something more than the expression of experimental failures to separate and distinguish.

Hitherto the terms have been employed to express matters of fact rather than of theory. These groups have been found experimentally to be chemically non-separable among themselves, and to possess a chemical character which apparently at least is identical. Nothing in this description necessarily implies that for all time they will remain chemically non-separable, or that some day slight differences of chemical character will not be discovered. The case of the rare-earths has often been quoted in this connection,[7] but a fuller examination shows that there is no real analogy between the cases. In the first place, among the rare-earths the difficulty is probably less in their separation, than to find means for determining what the effect has been of any attempt at separation. The only general test is the equivalent, and as the equivalents are close together the test is a very coarse one, so that separation of the constituents must already have proceeded to a considerable extent before it can be ascertained by such means. When, as in the case of praseodymium and neodymium, optical methods, such as the absorption spectrum, give an independent means of recognising the separate constituents of the mixture, the task of chemical separation is enormously simplified, whereas in the radio-elements extremely delicate and sensitive quantitative methods can always be devised to detect the least change in the ratio of a pair of non-separables in a mixture. The position with regard to the radio-elements is that more than one constituent is known to be there and yet they cannot be separated, whereas with regard to the rare-earths it is not known how many constituents are present until after long and laborious attempts at separation. There can be no question that the non-separables among the radio-elements, if they are not absolutely identical in every respect chemically, yet appear to be so at the present time, and that the constituents are certainly much more nearly alike than in any known case. It must not be forgotten, also, that a successful separation in some cases, as, for example, radio-lead from lead, would be very profitable, technically.

But now we may go on to consider the reasons underlying these facts, which make it appear that chemical non-separability and identity may be something more than merely the present limitations of the art of chemical analysis. It is necessary to reflect that, apart from radioactive evidence, the existence of more than one element in, for example, uranium or thorium could not have been suspected, to realise that chemical analysis, in general, is really not an analysis of matter into homogeneous elements at all, but rather into a number of types, homogeneous merely in their chemical behaviour. Each type is recognisably different from other types by chemical methods, but may or may not be homogeneous as regards, for example, other properties, such as atomic mass and stability. The periodic law expresses a *per saltum* rather than a gradual change of chemical properties. There are a limited number only of possible chemical types represented by the separate places in the Periodic Table. But these types in the only cases, those of the radio-elements, where we have an entirely independent method of analysis, prove *not* to be homogeneous.

It will be a great convenience if some new word is employed to express this newly revealed complexity of matter. The words "isotope" and "isotopic" suggest themselves. Henceforward a group of two or more elements occupying the same place in the Periodic Table, and being in consequence chemically non-separable and identical, will be referred to as a group of isotopes, and, within the group, the separate members will be referred to as isotopic. Thus ionium, thorium, and radio-thorium are isotopes, and mesothorium I is isotopic with radium.

When an α-particle is expelled it carries with it two

atomic charges of positive electricity, and the expulsion of these two positive charges from the atom affects the valency of the product, as Fajans has pointed out, exactly as in ordinary electro-chemical changes of valency. If the atom were initially in group IV, for example, its ion is tetravalent, and carries four atomic charges of positive electricity. Two such charges having been expelled with the α-particle the product is in the divalent group II non-separable from radium. The mass in this case is four units less. So with the β-ray change. The β-particle is a negative electron, and the loss of this single atomic charge of negative electricity increases the positive valency of the product by one. Radium B, for example, isotopic with lead, expels a β-particle and becomes radium C, isotopic with bismuth. But the mass in this case is practically unchanged. When one α- and two β-particles are successively expelled in any order the mass is reduced by four units, but the electrical content, as defined by the relative number of positive and negative charges in the atom, is unchanged. Hence the place occupied by the atom is the same as initially, and two atoms of different mass come to have identical chemical character, that is, are isotopic. But isotopes are not confined to the members of one disintegration series. *Whenever* two or more radio-elements fall into the same place in the Periodic Table, then, independently of all considerations as to the atomic mass, the nature of the parent element, and the sequence of changes in which they result, the elements in question are chemically non-separable and identical. As will later appear, this identity probably extends also to most of the physical properties, such as volatility and spectrum reactions.

The place occupied by the atom in the Periodic Table is thus primarily not a function of its mass, as has hitherto been supposed, but of its electrical content, as defined, and only to a secondary extent of its mass. In consequence a chemical element is not necessarily homogeneous, and its atomic weight may be, and possibly, in general, is, a mean value rather than a natural constant. In the vertical rows, on the other hand, where the mass varies by large steps as between the various members of the same family of elements, the members are analogous but not identical in chemical properties. Thus the Periodic Law represents the chemical character of matter as the function of two variables rather than of one. Of these two variables the electrical content is the essential variable in the horizontal columns, and the mass in the vertical columns. As between the successive members of the horizontal rows, the mass, whilst without direct influence on the chemical properties, is indirectly of paramount importance because of its effect on the stability of the atoms. If we exclude the radio-elements and confine ourselves only to the stable or common elements, for which the Periodic Table was originally constructed, it is clear that, whether a given element is homogeneous or not, is, in the end, a question whether in that place in the Periodic Table more than one isotope is stable—that is, whether atoms of different mass can permanently exist as stable systems when in association with the particular charge necessary to give the chemical character of the particular element in question. If only one isotope is stable the element must with lapse of time become homogeneous, unless, as in radioactive changes, the unstable isotope is being continuously maintained. But it may not even be the general rule that only one isotope is stable. Considerations of this kind may explain the "exceptions," such as argon and tellurium, to the Periodic Law, and the absence of mathematical relationships between the atomic weights.

Lastly, it appears likely, though the evidence is still incomplete, that different elements occupying the same place in the Periodic Table will be indistinguishable in spectrum and many other physical properties. This, at first sight, rather startling deduction is in accord with modern changes that have taken place with regard to our ideas of the origin of spectra. It is no longer possible, as was at one time supposed, to regard each line in the spectrum as due to the particular vibration of a single electron, if only for the reason that there are now known to be relatively few electrons in the atom, probably not more than half the number representing the atomic weight of the element in terms of hydrogen. The spectra, in all probability, originate from the accelerations of only a very few electrons in their orbits in an outer ring around the atoms, the same few electrons as condition chemical valency and affinity, and the great majority of the physical properties also. The periods of these orbits are fixed by laws formally identical with those of Kepler for the length of the planetary year, in which electrical action, varying inversely as the square of the distance, takes the place of gravitational action. Again, it is the electrical content rather than the mass which fixes the periods of these electrons and the wave-length of the lines in the spectrum. As the chemical character of a group of isotopes is not affected by variations of mass, it is to be expected that their electromic systems, which condition the chemical and spectroscopic properties alike, should be identical, and therefore that all the elements occupying one place in the Periodic Table should give the same spectra. Thus, apart from questions of stability, which operate only in the case of the radio-elements, the only properties available for distinguishing between the individuals of a group of isotopes are those which depend directly upon the mass, and these are relatively few. In the case of gases, Sir J. J. Thomson's new positive ray method of gas analysis as a means of distinguishing, and diffusion as a means of separation, have already achieved a notable success in this direction in the case of neon. This gas appears to be a mixture of two isotopes in very unequal proportions, possessing the same spectrum, the constituents differing in atomic mass by two units, and being non-separable by such methods as fractionation from cold charcoal, but separable by diffusion.

THE PERIODIC TABLE

FIG. 2 gives a common method of representing the Periodic Table, which, whilst very convenient, is misleading, in that it does not properly represent the continuity of the arrangement, as, for example, was so well shown in 1886 by the continuous spiral "figure of eight" of Sir William Crookes.[8] It is necessary, therefore, to ask first how the ends of each horizontal column are joined respectively to those of the preceding and succeeding columns. For a casual inspection might cause one to think that there were vacant places—for example, between xenon and iodine, or silver and palladium—which do not in fact exist, but which appear in the diagram, owing to the necessities of the two-dimensional mode of representation adopted. Such an interpolation of an argon type of gas between silver and palladium, or copper and nickel, or of an eighth group of elements between xenon and iodine, or krypton and bromine, would be absurd, and this leads one to recognise an important point overlooked in the ordinary presentation. The passage from the end of the A horizontal rows to the commencement of the next B row occurs through the eighth group, where the variation of chemical properties from place to place is at a minimum, whereas the passage from the end of the B row to the beginning of the next A row takes place with the maximum variation of chemical properties between the succeeding elements.

A diagrammatic representation of this has been attempted in Fig. 3, which may be regarded as a "figure of eight" arrangement brought up to date. The sharp bend at the argon members and the straight lines at the eighth group members indicate the differences in the rate of change of properties in the passage from one place to the next. It is of course a mere diagram, to represent clearly that the elements between { } in Fig. 2, which may be called the "interpolated elements," and between [ ], the rare-earth

PERIODIC TABLE OF THE ELEMENTS

	Group O.	Group I.	Group II.	Group III.	Group IV.	Group V.	Group VI.	Group VII.		Grou
Hydrogen 1.008	Helium He 3.99	Lithium Li 6.94	Beryllium Be 9.1	Boron B 11.0	Carbon C 12.00	Nitrogen N 14.01	Oxygen O 16.00	Fluorine F 19.0		
	Neon Ne 20.2	Sodium Na 23.00	Magnesium Mg 24.32	Aluminium Al 27.1	Silicon Si 28.3	Phosphorus P 31.04	Sulphur S 32.07	Chlorine Cl 35.46		
A	Argon A 39.88	Potassium K 39.10	Calcium Ca 40.07	Scandium Sc 44.1	Titanium Ti 48.1	{ Vanadium V 51.0	Chromium Cr 52.0	Manganese Mn 54.93	Iron Fe 55.	
B		Copper Cu 63.57	Zinc Zn 65.37	Gallium Ga 69.9	Germanium Ge 72.5 }	Arsenic As 74.96	Selenium Se 79.2	Bromine Br 79.92		
A	Krypton Kr 82.92	Rubidium Rb 85.45	Strontium Sr 87.63	Yttrium Yt 89.0	Zirconium Zr 90.6	{ Niobium Nb 93.5	Molybdenum Mo 96.0	—		
B		Silver Ag 107.88	Cadmium Cd 112.40	Indium In 114.8	Tin Sn 119.0 }	Antimony Sb 120.2	Tellurium Te 127.5	Iodine I 126.92		
A	Xenon Xe 130.2	Caesium Cs 132.81	Barium Ba 137.37		[Lanthanum La 139.0	Cerium Ce 140.25	Praesodymium Pr 140.6		Neodymium Nd 144.3	Samari Sa 150
	Europium Eu 152.0		Gadolinium Gd 157.3		Terbium Tb 159.2	Dysprosium Dy 162.5		Erbium Er 167.7		
		Thulium Tm 168.5	Ytterbium Yb 172.0	Lutecium Lu 174.0]		{ Tantalum Ta 181.5	Tungsten W 184.0	—	Osmiu Os 19c	
B		Gold Au 197.2	Mercury Hg 200.6	Thallium Tl 204.0	Lead Pb 207.10 }	Bismuth Bi 208.0	(Polonium)	—		
A	Radium Emanation 222.	—	Radium Ra 226.0	Actinium	Thorium Th 232.4	Uranium X_2 (Brevium)	Uranium U 238.5			

Only the four spaces marked are vacan places.

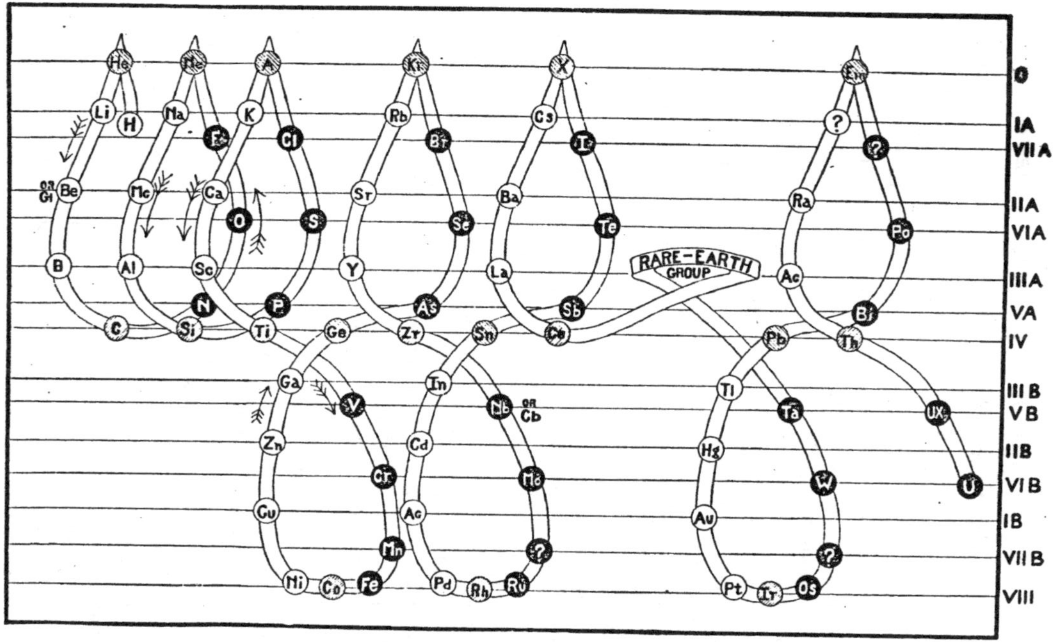

Electro-Positive Elements, above plane of paper, black letters on white ground.
Electro-Negative Elements, below plane of paper, white letters on black ground.
Intermediate Elements, in plane of paper, black letters on sectioned ground.

FIG. 3.

group, depart in their behaviour from the course followed by the two "short" periods and the first part of the first "long" period at the beginning of the table. In the former, the rate of change of properties from place to place diminishes continuously to the minimum at the eighth group as approached from either side, instead of, as before, increasing to a maximum at the zero group. In the latter, the rare-earth group, the rate of change is very slight throughout the whole group, and moreover, as the name indicates, all of these elements appear to belong to the trivalent family of the earth elements, of which aluminium is the commonest representative. But in spite of this the normal course is resumed when tantalum is reached, just as if it had never been interrupted. It is this latter part of the table with which we are more immediately concerned. Any theory of the nature of the elements which cannot account for these well-marked characteristics is not of much use in chemistry, and no such theory has yet been attempted. But it is well to remember that what has to be explained is now an extremely definite, if puzzling, relationship.

In Fig. 3, elements lettered black on white background are supposed to lie above the plane of the paper, those lettered black on a cross-hatched background in the plane, and those lettered white on a black ground, below the plane. The first are the electro-positive or basic elements, the second are the intermediate or neutral elements, and the third are the electro-negative or acidic elements.

It will be noticed that, in Fig. 3, the change from the A to the B series and *vice versa* is regarded as taking place in the middle of the series at the members of group IV., and not at the end of the series, as in Fig. 2. Thus antimony and tellurium are regarded as belonging to the A rather than the B series. In this book the lettering of Fig. 2 will be followed for reasons of convenience, although it is undoubtedly inferior to that of Fig. 3.

CHEMICAL AND ELECTRO-CHEMICAL ADVANCES

IN 1911 the chemistry of only some twelve of the radio-elements was definitely known. A direct systematic investigation of those still indefinite, in which they were studied each in turn with a view to finding out (1) whether they resembled chemically any known element, and (2) if so, whether they were chemically identical with, and non-separable from that element, as in so many previous cases, occupied Fleck two years.[1] The results may be condensed into very few words. It appeared early in the work that polonium was the only radio-element possessing a chemical character to itself unshared by any other, and even polonium, as we know now, is only apparently an exception, because its isotopes are all short-lived. Uranium X_1 was found to be chemically non-separable from thorium, mesothorium II from actinium, radio-actinium from thorium, the three B members of the active deposit groups from lead, the three C members from bismuth, and radium E from bismuth. This is a fine set of purely experimental results, undertaken deliberately without any hypothesis, in order that the data might be available for an adequate discussion of the question. The members still unelucidated, at the time the underlying law revealed itself, were the three A members, the hypothetical C′ members, of excessively short life, giving the long-range α-rays of thorium C and radium C, actinium D, thorium D, and radium C_2. It was predicted from the generalisation that the A members and the C′ members should be isotopic with polonium, and that actinium D, thorium D, and radium C_2 should be isotopic with thallium. Fleck was able to verify these predictions in the case of radium A, thorium D, and actinium D. The two latter may be quantitatively separated from the active deposits by precipitating potassium chloroplatinate in the solution, the chloroplatinate of thallium being the least soluble chloroplatinate known. In their other reactions they resemble thallium completely.

The following table of isotopes includes all the cases it has so far been found possible experimentally to examine. All the elements in each line are isotopic with one another, and the numeral at the beginning of each line indicates the place occupied by the group in the last long period. The case of the two uraniums will be considered later (p. 17).

VI. A *Uranium I.*—Uranium II.
IV. A *Thorium.*—Uranium X, Ionium, Radio-thorium, Radio-actinium.
III. A *Actinium* —Mesothorium II.
II. A *Radium.*—Mesothorium I, Th ım X, Actinium X.
O. *Emanation* of Radium, Thorium, .d Actinium.
VI. B *Polonium* (or Radium F).—Ra .um A.
V. B *Bismuth.*—Radium C, Thorium C, Actinium C, Radium E.
IV. B *Lead.*—Radium B, Thorium B, Actinium B, Radium D.
III. B *Thallium.*—Thorium D, Actinium D.

Electro-chemical Advances.—Although less directly connected with the present subject, mention must be made of some notable electro-chemical advances by G. von Hevesy,[9] on account of their great intrinsic interest. He showed that the deposition of the active deposit products upon a metal placed in their solution depends primarily upon the potential difference between metal and solution, and that all the B members and all the C members, respectively, exhibit in this respect an identical behaviour. With copper and silver electrodes, where the metal is ordinarily positive to the solution, the C member is deposited in excess, and may even be so obtained pure, but, as the potential difference decreases and changes sign, more and more of the B member is deposited. At -0.6 volt, measured against the normal calomel electrode, the two members are deposited in equilibrium, whilst with greater negative potentials, as with metals, such as magnesium and zinc, for example, the B members are deposited in excess. Conversely, from the ratio of B to C deposited, the electrode potential, if unknown, may be deduced, and this furnished the starting point of numerous applications of the radio-elements as "indicators" in ordinary physico-chemical measurements.[10] This is a rapidly growing and fruitful development, but little can be said here beyond the general remark that, if an element has a radioactive isotope, almost any kind of quantitative determination with it may readily and with advantage be converted into a radioactive determination. For example, the solubility of lead sulphide and chromate, and the amount of lead chloride entrained in a silver chloride precipitate, may readily be determined, by adding radium D, from decayed radium emanation, to the lead salt, and estimating it electroscopically.

The question of the valency of the radio-elements was examined by the same author by measuring the rate at which the cathions of the radio-element diffuse in a solution containing a great excess of the anion. The results obtained confirmed the accepted views on the valency of the radio-elements, where these were known. In other cases the method at first gave values which were at variance with Fleck's results, but these anomalies have now been satisfactorily accounted for.[11] More recently it has been established that the chemical identity of isotopes extends also to their electro-chemical behaviour, whilst an extremely thorough and systematic examination has confirmed the complete chemical identity of radium D and lead.[11A]

From a review of the electro-chemical evidence Fajans put forward the generalisation that the product of an α-ray change was more electro-positive than the parent, whilst that of a β-ray change was more electro-negative. In conjunction with the then available chemical evidence this led him independently to the complete periodic law generalisation.[4] The electro-chemical reasoning suggested the first method of separating uranium X_1 from uranium X_2. Anticipating, the latter, being formed in a β-ray change, should be more easily deposited on metals than the parent, and was in fact first separated from the solution of the mixture by deposition upon lead (p. 19).

THE CONNECTION BETWEEN THE SEQUENCE OF CHANGES AND THE CHEMICAL PROPERTIES OF THE PRODUCTS

IN the expulsion of an α-particle the atom loses four units of mass and two atomic charges of positive electricity, and in the expulsion of the β-particle the atom remains of practically unchanged mass, but loses a single atomic charge of negative electricity, or an electron, the mass of which may be considered negligible. The plate (Fig. 1) depicts the passage of the radio-elements in the three disintegration series through the Periodic Table. The ordinates represent units of atomic mass, and the abscissæ the number of successive places in the Periodic Table, which correspond, as we now know, to successive changes of the electrical content of the atom by one unit of charge. In passing from right to left of the diagram each place corresponds with an increase in the electrical content of the atom by one unit of negative electricity, or with a decrease of one unit of positive electricity. The numbers of the Families—for example, I for the alkali-metals, II for the alkaline-earth-metals, and so on—correspond well with the usual number of free positive charges carried by the atom when in the form of a cathion. Starting from the group of zero valency, say, from argon, a step to the right into group I., A gives us an element, such as potassium, the ion of which carries a single positive charge, whilst a step to the left into group VII., B gives us an element, such as chlorine, the ion of which carries a single negative charge. In the latter case, however, as Abegg has pointed out, we must consider both "positive" and "negative" valencies, and the arithmetical sum of these is always eight, the positive valency being represented by the group number, as, for example, in Cl_2O_7, and the negative valency by eight minus the group number, as in KCl. These considerations appear to apply equally to the part of the Periodic Table with which we are directly concerned; though, as is well understood, increase of atomic weight tends to accentuate the electro-positive character, and to cause the element to exhibit chiefly its positive rather than its negative valency. What holds good in the case of the VIIIth and neighbouring groups is open to conjecture.

In Fig. 1, in conformity with the rules already set forth, an α-ray change is depicted by a long arrow bridging over from one place to the next but one, pointing from right to left and sloping upwards to represent the change of mass, whilst the β-ray change is depicted as a short horizontal arrow pointing from left to right, and only bridging the distance from one place to the next. Owing to the absence of all direct information with regard to the atomic weight of actinium and its products, and of the parentage of this element, the actinium series is shown below the others in a separate part of the diagram. It appears likely that the atomic weights of actinium and radium and of their corresponding products are identical.[12]

The plate may now be systematically discussed with reference principally to the advances that have been made since the first part of this book was written, though, as already mentioned, most of the essential advances were made independently of any theoretical considerations before the generalisation in question was discovered.

We may consider first the case of thorium and actinium as far as the emanations in the zero group. The new data are the isotopism of mesothorium II and actinium, and of radio-actinium and thorium. In each case the rules are perfectly obeyed, as shown below, and they need not be considered in detail (Fig. 4).

The two rayless changes of mesothorium I and actinium, the only ones remaining in which some radiation has not yet been detected, obey the β-change rule, and it is probable that a very feeble β-ray is expelled in these cases also.

But in the case of the uranium series two advances of considerable importance call first for consideration. Boltwood showed from an analysis of the way in which the α-radiation of uranium minerals was distributed among the various products, that uranium gave two α-particles for every one given by each of the other six α-ray giving members of the series (I, p. 37). This is most naturally explained if ordinary uranium consists of a mixture of two chemically non-separable elements in the same series, each giving one α-particle per atom disintegrating. A careful examination showed that there are two sets of α-particles in the uranium α-radiation, with ranges in air at 15° and atmospheric pressure, 2.5 and 2.9 cms. respectively.[13] Further, from the Geiger-Nuttall relation (I, p. 9), it can

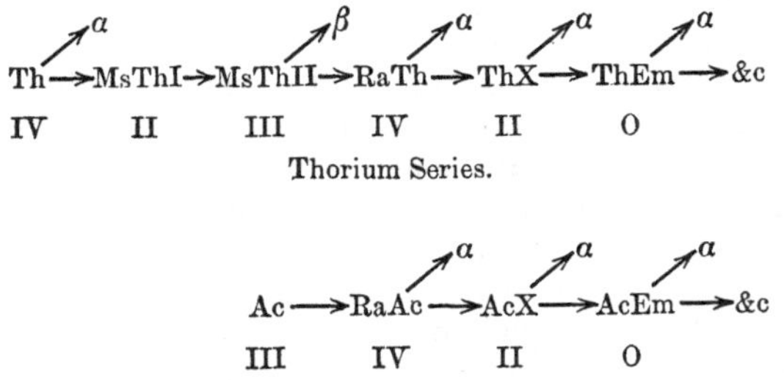

Actinium Series.

Fig. 4.

be calculated that if the 2.5 cm. range corresponds with the uranium, *i.e.* uranium I, for which the period of average life is 8×10^9 years, the period of the other uranium, uranium II, calculated from the range 2.9 cm. of its α-particle, is about 3×10^6 years, whilst that of ionium on the new data is about 3×10^5 years. The relative quantities of uranium I and II being proportional to these periods, there would be thus about 1 milligram of uranium II per gram of uranium. If its atomic weight is assumed to be less than that of uranium I by 4 units, the accepted value for the latter will only be affected in the third decimal place.

The question now arises whether uranium X is the product of uranium I or of uranium II. It follows at once from the analogy in the thorium series, in which similar non-separable pairs—thorium and radio-thorium, mesothorium and thorium X—are not successive, but are separated by members of distinct chemical nature, that in all probability uranium X is the product of uranium I and the parent, rather than the product, of uranium II. Thus if we write:

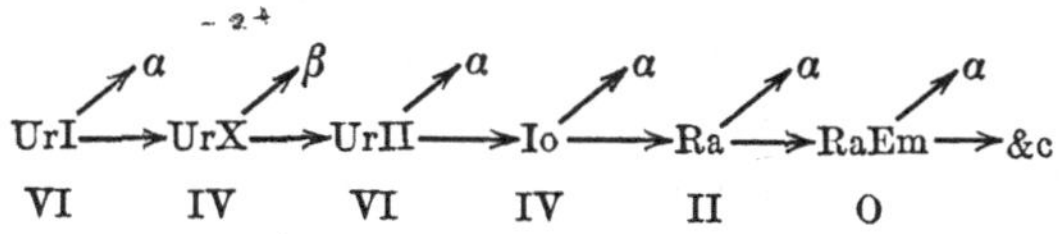

Fig.5. Incomplete Uranium Series.

we shall have another case of the alternations already familiar, the two uraniums on the one hand and uranium X and ionium on the other being respectively pairs of non-separables, analogous to thorium and radio-thorium, and to mesothorium and thorium X. This was pointed out by numerous authors simultaneously.

But clearly, if the rules hold good, one β-change is here missing between uranium I and uranium II. The loss of a β-particle by uranium X should result in a product in group V. A, to which the nearest analogue is tantalum, and no representative of this place was previously known. But uranium X expels two sets of β-rays, one hard and one soft (I, p. 38), and an examination by Fajans and Göhring established that these radiations are produced in successive changes rather than in a single change.[14] The two products are therefore known as uranium X_1 and uranium X_2. Uranium X_2, when discovered, was found to be a very short-lived radio-element, with period of average life about 1.6 minutes, and giving only the hard β-rays (presumably the γ-rays also) of uranium X. Uranium X_1, on the other hand, the substance with period 35.5 days, gives only the soft β-rays, and produces uranium X_2. The two may be separated by several methods. But the reactions which show best its chemical character also show that chemically it is allied to tantalum. Thus when uranium X separated from uranium along with a trace of thorium, is added in acid solution to an alkaline solution of potassium hexatantalate, uranium X_2 is precipitated almost quantitatively with the tantalic acid. Again, when uranium X in solution is passed through a filter on which has been spread a few milligrams of moist tantalic acid, the uranium X_2 is left behind to a great extent on the filter. When lead is precipitated as chloride in a uranium X solution, the precipitate contains excess of uranium X_2, whereas, with lead sulphate as precipitant, the converse is the case.

Since uranium X_2 occupies a place to itself in the Periodic Table, which formerly was vacant, it constitutes a new chemical type, and therefore should be given a distinctive name. The suitable name "Brevium" has been chosen by its discoverers.

Hence the uranium radium series so far as known is also in complete accord with the generalisation (Fig. 6).

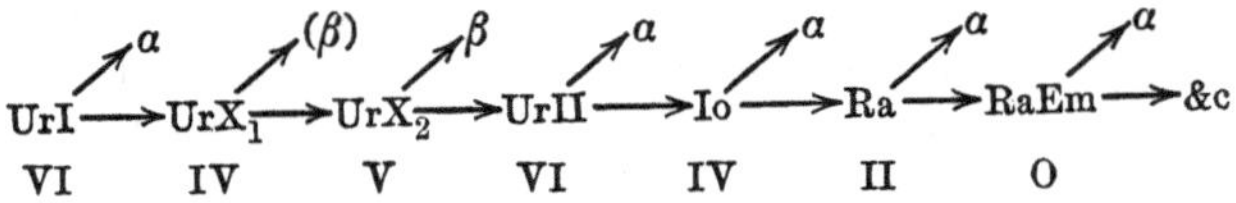

Fig.6. Complete Uranium Series.

Between uranium and the zero group every place is now occupied, except that of the alkali-metal of group I., which is still vacant. Six places accommodate no less than seventeen elements, instead of the two, uranium and thorium, previously recognised. It is interesting to note that only the first new element to be found occupying a vacant place would be chemically recognised, however many existed. Thus there never has been any doubt as to the new elementary character of radium, actinium, and polonium. The places they occupy were previously vacant. But as regards, for example, ionium or mesothorium I, though the elements when discovered were also new, their chemical properties were not new, but were the same as those of others already perfectly well known. It is clear that the chemical properties alone, in absence of any other evidence, are a very insufficient criterion on which to judge of the homogeneity of any material.

THE BRANCHING OF THE DISINTEGRATION SERIES

BEFORE we can consider the disintegration series beyond the emanation group, some considerable changes that have recently been made must first be dealt with. It has long been recognised (I, pp. 57 and 79) that the transformations of the active deposits were not completely understood. The first step was the discovery of the evanescent A members of the thorium and actinium series corresponding with radium A. Their separate recognition, since previously their α-rays had been attributed to the emanations, removed one of the grounds for supposing (I, p. 37) that in these series the different α-ray changes did not all produce one α-particle per atom disintegrating, as in the case of the uranium series.[15]

The numbers of α-particles expelled in the thorium series were directly counted by Barratt, who showed that, in equilibrium, per 100 α-particles of thorium emanation of range 5.0 cm. (at N.T.P.), 100 of range 5.4 cm. are given by the A member, whereas 35 of range 4.8 cm. and 65 of range 8.6 cm. are given by thorium C.[16] This shows directly and beyond doubt that the series branches at or before the C member, 35 % of the atoms going by one route and 65 % by another. The numbers of the atoms going along the two branches being of the same order makes the series comparatively simple to unravel experimentally. As a matter of fact, however, the far more difficult case of radium C, investigated by Fajans, was the first example of branching to be definitely established.[17] It is clear (I, p. 58) that radium D must be the direct product of radium C in its α-ray change, because it is obtained from it in relatively large amount by recoil methods. Hence radium C_2, which gives β-rays and no α-rays, cannot be intermediate as represented on I, p. 55. Moreover, the small relative quantity of radium C_2 obtained by recoil, the β-activity of which is only $\frac{1}{20000}$th of that of the source, cannot be explained, as it was formerly, by supposing that it is produced in a β-ray change. Because, in other β-ray changes, for example that of radium E, no trace of the product recoils. That the relative quantity of radium C_2 is really, as well as apparently, minute is shown by its being without influence on the growth of β-rays from pure radium B.[18]

Let us consider what theory would lead us to expect in the case of a branching of the disintegration series.[19] If an atom has a choice of modes of disintegration, it is most reasonable to suppose that the chance, in any small time-interval, that the atom will disintegrate will be for *each* mode fixed, as in the case of a simple change, independently of every circumstance whatever. In particular, each mode of disintegration will proceed, as in a simple disintegration, as though the other mode did not occur. If the radioactive constants for the two modes are λ_A and λ_B respectively, the differential equation expressing the change is:

$$dQ/dt = \lambda_A Q + \lambda_B Q \text{ or } Q(\lambda_A + \lambda_B),$$

and the quantity of substance diminishes exponentially, as in the case of a simple change:

$$Q_t/Q_o = \epsilon^{-(\lambda_A + \lambda_B)t}.$$

If the λ_A mode is supposed to be accompanied by the expulsion of α-particles, and the λ_B mode by the expulsion of β-particles, then *both* α- and β-rays will decay exponentially with the $1/(\lambda_A + \lambda_B)$ period. The proportion in which the atoms will distribute themselves along the two branches will be as λ_A and λ_B, respectively, are to $\lambda_A + \lambda_B$. We see at once that if λ_A and λ_B are of the same order, as happens to be the case in the thorium series, the two branches will claim similar proportions. But that if, as is the case with radium C, and as is more likely to be the general case, λ_A and λ_B are widely different, one branch will claim nearly the whole of the disintegrating atoms, and the other only a small fraction.

We have seen that the thorium series branches at or before the C member. If branching occurs at thorium C, that substance will *not* be capable of separation into two distinct products, one giving the α-rays of range 4.8 cm., and the other those of range 8.6 cm. Whereas, if it occurs before thorium C, at, say, thorium B, then thorium C might be expected to be resolvable into two products, and, moreover, these two products *must have the same period of change*, otherwise the complex α-radiation of thorium C could not decay exponentially as it does. All attempts to separate two products from thorium C have proved unsuccessful, no alteration in the 35 : 65 ratio of the numbers of the two sets of α-particles expelled has been effected or observed under any circumstances—for example, when the α-rays are growing or decaying—and the conclusion may be considered established that thorium C is a homogeneous member, disintegrating dually.[20] Other conclusions have since been shown to be due to secondary disturbing causes.[21]

As to the distribution of the various radiations in the active deposit of thorium, it has been shown that, although all the γ-rays, only a portion, and that the smaller and less penetrating portion, of β-rays come from thorium D, the greater portion coming from thorium C. The thickness of aluminium required to cut down the β-rays to half value is, for the rays from thorium D, 0.32 mm., and for those from

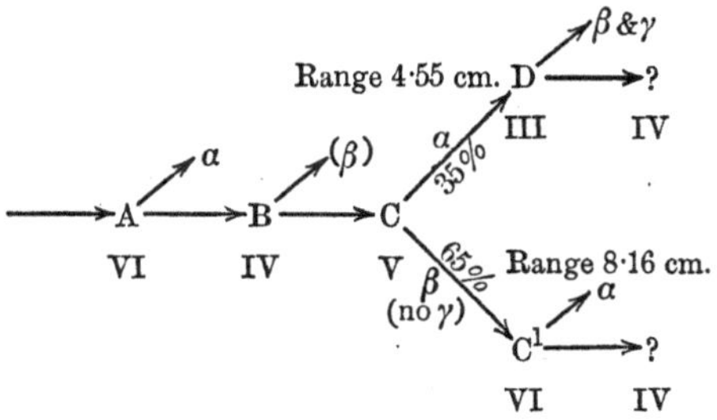

Fig.7. Thorium Series.

thorium C, 0.48 mm. This is a most interesting example of hard β-rays being unaccompanied by γ-rays, and of softer β-rays accompanied by the hardest γ-rays known, and is additional evidence that the two types of rays are not so simply connected as was at one time supposed.

The diagrammatic representation of these results is shown in Fig. 7. Thorium C disintegrates dually. The period of the substance, $(1(\lambda_A + \lambda_B))$, 79 minutes, being the sum of two separate periods in the ratio of 35 : 65, corresponds therefore with $1/\lambda_A$, 225.7, and $1/\lambda_B$, 121.5 minutes. In the quicker of the two simultaneous changes β-rays are expelled first, and then α-rays of 8.6 cm. range from an extremely short-lived product, designated as "thorium C′." From the Geiger-Nuttall relation, it follows that the period of thorium C′ can only be of the order 10^{-11} second, and hence it is experimentally indistinguishable as a separate product. In the slower of the two simultaneous changes, α-rays are first expelled, of range 4.55 cm., which corresponds well with the true period of this change, 225.7 minutes. The product is thorium D, which expels β- and γ-rays. It will be seen that in the one branch an α-ray change is followed by a β-ray change, and in the other a β-ray change is followed by an α-ray change.

Radium C.—An exactly analogous series of changes may be constructed for radium C, the difference being that here the two periods, instead of being in the ratio 35 : 65, are as about one to three thousand. In consequence the main branch claims the overwhelming proportion of the whole,

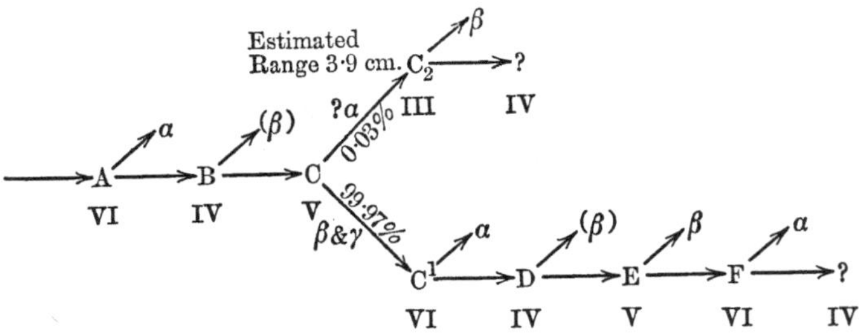

Fig.8. Radium Series.

the side branch in which radium C_2 occurs, only 0.03%. Hence the period of the main change is indistinguishable from that usually given for radium C, viz. 28.1 minutes, whilst that of the other is about 6.5 days. Naturally the latter period is at present only very roughly known (Fig. 8).

Here, as before, the faster of the two simultaneous changes is accompanied by the expulsion of a β-ray, but in this case γ-rays, also of a very penetrating character, are expelled. The product of the β-range change, as before, expels α-rays of long range, the range corresponding with an estimated period of 10^{-6} second. The product of this change is not, as in the thorium series, the unknown end product of the series, but the well-known radium D, as shown by the recoil experiments. Radium D is thus not at all analogous to thorium D. Now consider the slower of the two simultaneous changes. The product of this minor branch must be radium C_2, which gives β-rays, and, if analogy holds good, it should be produced in an α-ray change of radium C. The estimated period of the change is 6.5 days, so that the α-rays have a calculated range of about 3.9 cm., similar to those expelled by radium emanation. However, these have not yet been detected. It is an extremely difficult task to look for one α-particle of low range among 3000 of longer range, and some attempts made by the writer to find them led to no result. Assuming, however, that the scheme is correct, radium C_2 should be analogous to thorium D, both being β-ray giving products formed in the α-mode of the dual disintegration. The nomenclature is necessarily terribly confusing, each product, as found, having been named by the discoverer, according to no complete or previously known scheme. But it must serve its purpose a little longer, until now that the whole of the series have been definitely brought under law and order, a complete and satisfactory scheme can be devised. It seems "best to bear the ills we have, than fly to others that we know not of."

Actinium C.—It now remains to discuss the actinium series. Earlier indications (I, p. 87) that the α-radiation of actinium C were also complex, the ranges of the sets, being very similar, have not been confirmed. We shall not go far wrong if at first we assume that, if actinium C does disintegrate dually, the proportion of the atoms travelling along the minor branch is so small as to be practically negligible in a first examination. Which branch is it, the α- or β-mode of disintegration, which actinium C mainly follows? The range of the α-rays of actinium C (5.12 cm. at N.T.P.) obviously corresponds with the period (3.1 minutes). Hence *the main branch is here analogous to the minor branch* in both the thorium and radium series, and the product actinium D, giving both β- and γ-rays, is analogous to thorium D and radium C_2. The series is thus quite straightforward (Fig. 9).

Recent evidence of a small proportion, some 1 in 600, of faster α-rays of range about 6.1 cm. has, however, been

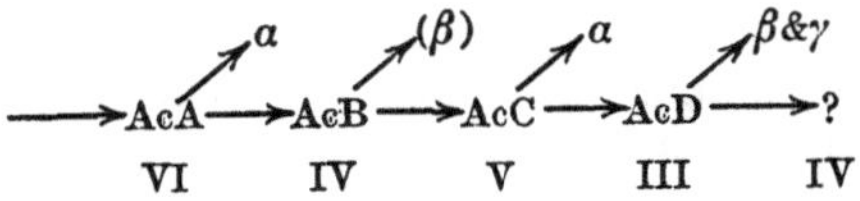

Fig. 9. Main Actinium Series.

obtained.[22] If this is confirmed it means that actinium C must disintegrate according to a second slower β-ray giving mode, with separate period about ten hours to produce a product giving α-rays and having a period estimated from their range, which is almost exactly that of actinium A, of about 0.003 second (Fig. 10). It is curious to note that the α-ray mode of radium C has a calculated period almost the same as that of the radium emanation, and no doubt

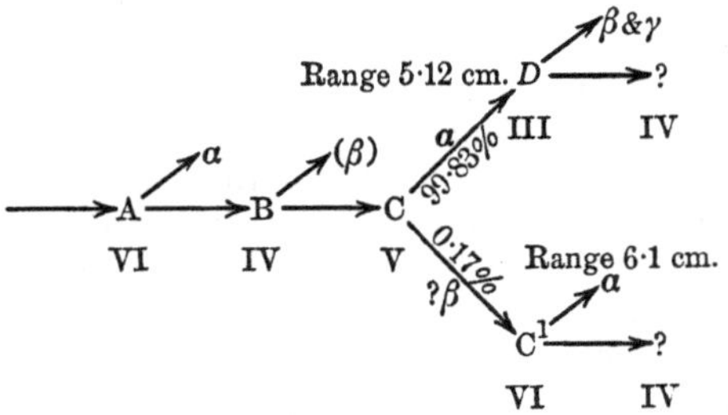

Fig. 10. Probable Actinium Series.

there is method underlying these periods. But as the actinium active deposit, ordinarily prepared, always contains a minute amount of recoiled actinium X, it is best to wait for the full results before accepting the scheme of Fig. 10.

Confirmatory evidence of the correctness of the above schemes has recently been obtained by determining the range of the recoil products of thorium C and actinium C. As Wertenstein showed, the recoil products of an α-ray change have ranges in air and hydrogen proportional to the ranges of the α-rays expelled in the same change, and this, of course, is to be expected.[23] The ranges are small, some 0.14 mm. of air at N.T.P. for radium D, but are usually determined in air at low pressure. Now the range of the recoil products of thorium C and actinium C, viz. thorium D and actinium D, are almost exactly the same, 0.12 mm., just as the ranges of the α-rays expelled are very similar. But thorium C gives a second inactive recoil product, from thorium C′, which, like that from radium F, is investigated, not by its activity, but by the intense ionisation it produces in its flight. This second recoil product has a range of 0.175 mm. of air.[24]

It will be seen that, in their present form, as shown (Figs. 4, 6, 7, 8, 9), the three disintegration series perfectly obey the periodic law generalisation. For the hopelessly unstable A and C′ members, all of which should fall in group VI. B, and be isotopic with polonium, we have direct evidence only in the case of radium A. The B and C members fall into group IV. B and V. B as required, and are chemically non-separable from lead and bismuth respectively. Actinium D, thorium D, and radium C_2 fall into group III. B, and in the case of the two former have been proved to be chemically non-separable from thallium. Radium D is non-separable from lead, as is required if radium C′ is in group VI. B, as supposed. Its product, radium E in group V. B, is non-separable from bismuth, whilst the penultimate member, radium F or polonium, is in group VI. B.

Finally, with regard to the nature of the six end products, it will be noticed with interest that they all fall in group IV. B, and therefore must be isotopic with lead. As regards radium, it has long been regarded as practically certain that its end product must be lead, though this still needs direct proof. But as regards the nature of the end products of thorium and actinium, no hint previously has been forthcoming. Recent experiments have been done on a preparation of radium D, derived from the change of 0.2 curie of radium emanation, and estimated from its age to contain 15 % of radium G, the final product. A micro-chemical test by means of the delicate reaction of lead nitrite with copper and potassium nitrites, in which a characteristic crystalline triple nitrite is formed, showed clearly the presence of lead in the product. But it is too early to judge whether it was due to the radium D or G, or to both products together.[25]

NATURE OF THE END PRODUCTS. ATOMIC WEIGHT OF LEAD

THE question thus arises whether common lead is a homogeneous element of atomic weight 207.1, as expressed in the International List, or a mixture of the end products of the three series in various proportions. There has long been an unexplained discrepancy between the atomic weight of lead and that calculated for the end product of the radium series. For the atomic weight of radium is now known with an accuracy at least as great as that of lead, owing to the splendid work of Hönigschmid with over one gram of pure radium chloride.[26] From his figure, 225.95, it is clear that the atomic weight may be taken without serious error as 226, so that the atomic weight of the end product, after the expulsion of 5 α-particles of mass four, would be 206, or 1.1 units below the value for lead. But the calculated value for both the end products of the thorium series is 232.4 minus 24, or 208.4. The end products of actinium would only be present in relatively small proportion, and may be neglected. If actinium has the same atomic weight as radium, both end products would have the atomic weight 210, which is the same as that of the end product of the minor branch of the radium series, formed in altogether negligible proportion. Hence it is clear that, whatever be the actual case, the atomic weight of lead agrees with what is to be expected for a mixture of the end products of the uranium and thorium series in similar proportion.

The difficulty at the present time is to be sure that what experimentally appear as the end products in all cases are really completely stable elements. It is quite possible that some of these products may be changing too slowly to give a detectable activity, and yet too rapidly to survive over geological epochs of time. Indeed, Fajans has suggested that the bismuth and thallium found in radioactive minerals may be the result of further changes of what have been regarded hitherto as the end products. These, for the thorium series, have the calculated atomic weight about 208, and can be supposed to pass by the loss of a β-ray into bismuth (208), or of a β- and α-ray into thallium (204). It is clear that atomic weight determinations of the lead from radioactive minerals, differing as widely as possible in their age and relative uranium and thorium content, would throw further light on the question. In this connection a generalisation of Fajans may be mentioned, which, whilst it does not hold without exception, holds probably in too many cases to be accidental.[12] He points out that in the α-ray giving members of an isotopic group the smaller the atomic weight the less stable is the element, whereas for the β-ray giving members the reverse is the case. Radium F is an exception in the first case. If actinium is assumed to have the same atomic weight as radium, its α-ray products obey the rule, except actinium X, which has a longer period than thorium X, the calculated atomic weight of which is two units greater. For the β-ray products actinium B, similarly, is shorter-lived than thorium B. Applied to the present case, it is to be expected that if the end product of thorium, analogous to radium D, resembles it in expelling a β-ray, it should be more stable than radium D (210), but less stable than radium G (206), because its atomic weight is about 208. On this view it is to be expected, therefore, that this product would not accumulate in minerals, but would slowly change into bismuth, or into an element isotopic with it. As a matter of fact, though it is easy to get uranium minerals from secondary rocks like the pitchblende of Joachimsthal, which are free from thorium, it is by no means easy to get thorium minerals free from uranium, and it is always possible that there is some genetic connection between the elements not yet disclosed.

It is of interest to note how nearly science has approached to the solution of the problem of the alchemist. If thallium could be made to expel an α-particle, or mercury one α- and one β-particle, the product would be isotopic with gold. Though, of course, this is not yet possible, there can be little doubt that success would follow the application of sufficiently great electric potentials, of the order of some millions of volts. So far as can be seen all insulating media, even a perfect vacuum, fail at potentials far short of this, conceivably by such disintegration of the material of the electrodes.[26A]

THE ORIGIN OF ACTINIUM

ALTHOUGH the long outstanding problem of the origin of actinium has not yet been cleared up, the number of possibilities to be considered is now greatly reduced, and it is to be expected that the answer to this question will soon be forthcoming. There is little doubt that actinium is a relatively short-lived radio-element, and that its period cannot be greater than a century at most. Mme. Curie, from direct observation, estimated the period provisionally as only thirty years (I, p. 83). From the fact also that intensely active preparations of it have been prepared, it scarcely can be a primary element, and the first question to be considered is its parent. On this point the rules give an alternative (Fig. 11). Either it must be formed from radium, or one of its isotopes in group II. A, in a β-ray change, or from "Eka-tantalum," group V. A, in an α-ray change. At first, before the discovery of uranium X_2, the writer suggested that the unknown product of uranium X, in group V. A, might disintegrate dually, like the C members in group V. B, giving uranium II in a β-ray, and actinium in an α-ray change.[5] This could only have remained unknown if the period of this unknown member were very long. In this case "Eka-tantalum" would constitute a new radio-element, obtainable from minerals in quantities sufficient for chemical examination. The discovery of uranium X_2 in this place, and its short life and absence of α-radiation, disproved this view.

An experimental examination of a preparation of radium bromide, containing 13.2 milligrams of radium (element), which was ten years old, and had not been subjected to any interference during that time, disproved in turn the second alternative.[27] No trace of actinium was found to be present in the preparation, although radio-thorium in distinct amount was obtained. The origin of this radio-thorium, presumably, is the minute amount of thorium which is known to be present in Joachimsthal pitchblende, from which the radium was extracted. Mesothorium I, isotopic with radium, would thus be present, and would produce radio-thorium. Since radio-actinium is isotopic with radio-thorium, the absence of any indication of actinium, though radio-thorium was detected, is a guarantee that no appreciable amount of actinium can have been formed. The conclusion was reached that actinium cannot be produced from radium

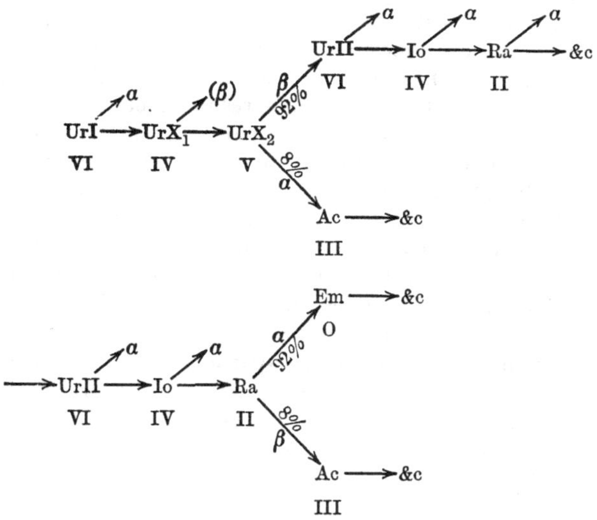

Fig.11. Alternative Possible Origins of Actinium. (Both now disproved).

unless it has a period of at least 15 million years, which is practically impossible, as this would mean that there must be about 170 grams of it per ton of uranium in minerals, and that the α-activity of pure actinium in equilibrium with its products could never be greater than 1650 times that of uranium.

It has been supposed that uranium X_1, rather than uranium X_2, might undergo the dual disintegration, in both modes, with expulsion of β-rays, producing two isotopes, therefore, in group V. A, one of which would be uranium X_2, and the other the missing parent of actinium.[28] Once, however, the possibility of a dual change is admitted, giving the same kind of rays in both modes of disintegration, other alternatives can be framed. It seems extremely likely that some such change does occur.

At present the most probable theory is connected with the existence of uranium Y, which Antonoff claimed to have discovered in 1911 (I, p. 40). The effect on which he based this conclusion was that the soft (β)-radiation of uranium X decays, in certain cases, abnormally rapidly for the first few days after separation from uranium. This effect has been called in question, but repeated and reaffirmed by Antonoff,[28A] and has recently been independently confirmed by the author so far as the main result is concerned. But if it is due to the existence of a separate product, uranium Y, this product appears now to be isotopic with uranium X_1, and to be formed directly from uranium, either I or II, presumably in a branch α-ray change. Antonoff states that the proportion of (β)-radiation contributed by uranium Y is of the right order to be expected if it is the first member of the actinium branch series, and this view also is independently supported. If uranium Y exists and is the parent of actinium, one α- and one β-particle must be expelled in the passage from group IV. A (uranium Y) to group III. A (actinium). So far there is, however, no evidence of the α-ray change.

THE SPECTRA OF ISOTOPES

OWING to the minute amounts of most of the radio-elements available, it has not yet been found possible directly to test the view whether isotopes possess identical spectra as well as identical chemical character. The view was put forward first upon the results of the spectroscopic investigation of two ionium-thorium preparations.[29] The clearing up of the series between uranium and radium makes definite the minimum estimate of the period of ionium, which was deduced from the absence of growth of radium from uranium, and which depends upon the absence of intermediate elements of long life between uranium and ionium. This minimum estimate of 100,000 years[30] agrees with the estimate of 300,000 years calculated from the range of the α-rays (p. 18). From this period it may be calculated that the ionium preparations spectroscopically examined must have contained at least 10 % by weight of ionium, and probably much more. Yet no new lines were observed in the spectrum, although Russell and Rossi found that an admixture of 1 % of ceria or uranium oxide with thoria could readily be spectroscopically detected.

The direct determination, for example, whether the spectra of thorium X and radium, or of radio-thorium and thorium, are identical, is not altogether impossible, but calls for large quantities of radioactive material, which is excessively costly.[7] Mme. Curie is engaged on the spectroscopic examination of radium D. On the view here supported, the spectrum of this element should, of course, be the same as that of lead. The recent generalisation of Prof. Hicks, that the magnitude of the atomic weight enters exactly into the series relations of spectra, is, however, strongly opposed to this view.

NEON AND METANEON

FROM an entirely independent field from that of radioactive change, and all the more welcome on that account, evidence has recently been forthcoming upon the questions of the homogeneity of chemical elements and the spectroscopic identity of isotopes. Sir Joseph Thomson has recently perfected an entirely new method for the determination of the masses of the positive ions in an ultra-rarefied gas under the action of the electric discharge—for the masses, that is, of the material particle constituting the so-called "positive-rays" or "canalstrahlen."[31] The method depends on the well-known principle due to the same discoverer of deflecting the trajectory of the ion by the simultaneous application of electric and magnetic fields. Both magnetic and electric forces act, in this case, in the same plane at right angles to the direction of the ion, and produce simultaneous deflections in two planes at right angles to this direction. In this way all the particles having the same ratio, e/m, of the charge to the mass, fall upon a photographic plate, placed at right angles to the initial direction of the ion, on a parabola, and from the number of parabolas the number of different kinds of ion in the gas may be at once found. From the parabola corresponding to the univalent hydrogen ion, which is found upon all plates, the value for e/m of the other ions may be deduced. From certain peculiarities in the photographic traces, for which the original paper may be consulted, singly charged ions may be distinguished from multiply charged ions, and thus the value of m found.

Several ions were so recognised, corresponding with no previously known particle. Some of these, notably "X_3," which is now regarded as H_3, the hydrogen analogue of ozone, are probably new molecular species, highly interesting in themselves, but not directly connected with the present subject.

In the case of the element neon, in addition to the known atom of mass about 20 a second was discovered of mass, about 22 which could not be ascribed to any known element. In consequence, an investigation of atmospheric neon was made by F. W. Aston, and a preliminary communication of the results was made at the 1913 British Association Meeting at Birmingham. A prolonged series of fractionations by the use of cold charcoal did not affect the density of the neon, or the relative amount of the heavier constituent, as shown by the positive ray method of analysis. Then a series of fractional diffusions, coupled with density determinations, was carried out, which resulted in a partial separation of the two constituents, as attested by a change of density. The elements appear to be identical in all their properties except atomic weight, and no change in the spectrum corresponding with the change of density has been observed. The continuation of these experiments, which, naturally, are long and laborious, should thus furnish valuable evidence on the questions discussed in this monograph, and particularly as to whether isotopes have the same physical properties apart from those depending directly upon molecular mass.

NATURE AND PROPERTIES OF ISOTOPES

ISOTOPES might be divided into three classes, distinguished by differences in one of three respects—atomic mass, stability, and internal atomic energy, though not necessarily by differences in the other two. Thus neon and metaneon, the corresponding members of the actinium and radium series, and the two end products of the thorium series, might be taken as examples respectively of the three classes, in which one variable only is known to differ. In the latter case, as pointed out by Fajans,[12] the two products are of the same atomic mass, and therefore, it might be argued from his generalisation, of the same stability; but they are formed from the same parent element with losses of dissimilar amounts of energy, and therefore, whether of the same stability or not, must be regarded as isotopic rather than identical, in derivation if not by the possession of any actual recognisable difference. The hitherto unsuspected possibility of the existence of the first class of isotopes very much confuses the significance to be attached to the meaning of the value of the atomic weight. What possible methods are available to test further the homogeneity of the elements?

It is clear that isotopism carries with it a material identity which extends far beyond the limits of chemical properties in the narrow sense. Thus all the members of the inert family are necessarily chemically non-separable and identical, in a purely negative sense, but are distinguished by different spectra, and can be separated readily by differences in volatility as well as in density. They cannot be considered isotopic. Neon and metaneon, on the other hand, distinguishable in density, are indistinguishable in volatility and spectrum.

Now, although it was not properly understood at the time, the very surprising observation was made ten years ago that the volatilisation temperature of the radium and thorium emanations were, within the error of the observations, identical.[32] It is true that differences manifest themselves in the progress of the condensation as the temperature is lowered, but these are not greater than can be ascribed to the enormous difference, 6000 to 1, in the concentrations of quantities of the two emanations producing similar radioactive effects, and to the effect of absorption by solid surfaces in raising the temperature of condensation, shown in so marked degree by charcoal. We must conclude that the electrical content of the atom rather than the mass is responsible for many physical properties as well as for the chemical nature of an element, and that isotopism carries with it a material identity in the case of these properties also. The possible methods of distinguishing between isotopes of the first class appear to be almost confined to the two methods already discussed, the new positive ray method and diffusion.

New atomic weight determinations, done on materials geologically as distinct and simple as possible, might furnish new evidence, but there is always the possibility that the different isotopes may be present in constant relative proportion—for example, as the end products of a branch series. Diffusion in solution is complicated, at least in electrolytes, by the probability that the diffusing particle is not the simple molecule but a much larger aggregate with solvent molecules. An attempt by the writer to separate the two uraniums in this way—a particularly favourable case—gave no result after diffusion had proceeded for a year. As Graham showed, rates of diffusion in solution do not, as in the case of gases, depend simply upon the magnitude of the molecular masses. But diffusion in the state of gas, in those cases where it is possible, ought to be effective. This opens a wide field for research, for volatile compounds of most of the elements exist, upon which diffusion and the positive ray method could be tried, even though the element itself is not volatile.

THE STRUCTURE OF ATOMS

ALTHOUGH physicists have not yet advanced their theories of atomic structure to the point of giving a reasonable explanation of the Periodic Law (Figs. 2 and 3)—and, indeed, have hardly got beyond the consideration of a few of the simplest elements—the advances dealt with here do to some extent contribute material data to the discussion of the nature of atomic structure. It is clear that the charges on the α- and β-particles do definitely come from the atom, and are not, as the writer once tried unsuccessfully to show in the case of the α-particle, derived subsequently to the disintegration, as by collision with the matter in the path of the ray. It is true that, in the best-known cases of α-ray disintegrations, three rather than two charges of positive electricity appear to be liberated from a system previously electrically neutral, one being found on the recoil atom.[33] But this may or may not be of significance, because the emission of a single δ-particle would be very difficult to detect, and might account for the third positive charge. But the shift of place of the atom in the Periodic Table is most naturally explained on the view that the charges expelled with the rays are intrinsic to the atomic structure.

Fajans[4] has argued that, since the changes of properties suffered in radioactive change are the same as those in electro-chemical change, radioactive changes must be concerned with the same part of the atomic system as ordinary chemical change, contrary to the accepted view. But this conclusion does not stand closer examination. Clearly upon it, for example, uranium in the uranous salts should be non-separable from thorium, so long as it were prevented from altering its tetravalency. There is a marked analogy in chemical properties between uranous and thorium salts which gives some colour to the supposition, but an examination, still unpublished, by Fleck and the writer, showed that the two classes of compounds are readily separable by chemical methods. The fair inference from this is that, as commonly supposed, chemical and electro-chemical properties are controlled by the outer ring of electrons, and that radioactive change is concerned with the nucleus. The expulsion of the helium nucleus, or helium atom with two positive charges, in an α-ray change, alters the electrical charge of the *nucleus*. The reduction of uranyl to uranous salts alters the electrical charge of the ring in an exactly equivalent way. Uranium in uranous salts is not isotopic with thorium, although the total electrical content of the atom is the same in each case.

Uranium X_1 loses two electrons as β-particles and becomes uranium II, whereas uranous salts lose two electrons in electro-chemical reductions and become the commoner uranyl compounds. Since the products are isotopic and the parents are not, this may be regarded as a definite experimental proof that the electrons lost in β-ray changes come from a different region of atomic structure than those lost in electro-chemical changes. The β-ray electrons like the α-particles must come from the nucleus.[34]

The occasional large deflections of α-rays in their passage through atoms led Rutherford to conclude that the mass of the atom must be concentrated in an exceedingly small central nucleus, in diameter only of the order of $\frac{1}{10000}$th of the diameter of the atom, and bearing a positive charge approximately one-half of the atomic mass.[35] Van der Broek has suggested that this positive charge, instead of being one-half of the atomic mass, might be the number corresponding with the place of the element in Mendelejeff's table, when the elements are arranged in sequence.[36]

As regards the sequence thallium to uranium, the suggestion may be regarded as proved by Fig. 1, so far as it concerns the relative value of the positive nuclear charge, and its variation by one unit from one place in the table to the next. If all the rare-earth elements are known (Fig. 2), the positive charge of the nucleus of the uranium atom would be 89. Since electrons, at least six in the case of uranium, corresponding with the six β-rays expelled in its disintegration, also exist in the nucleus, the positive charge must be at least 95. If the nucleus were made entirely of potential α-particles or helium atoms, for which four units of mass correspond with two units of charge, the positive eharge would be about 120. Whereas, if hydrogen is a second ultimate constituent of atomic structure, since for hydrogen unit mass corresponds with unit charge, the number would be still greater. It is clear, therefore, that in dealing with the nuclear charge we must distinguish between the algebraical and arithmetical sum of the opposite charges. It is the algebraic sum—that is, the difference between positive and negative—which fixes the position of the element in the Periodic Table. Isotopes of the first class might be defined as elements for which the algebraic nuclear charge is the same, but the arithmetical nuclear charge is different.

Chemical and electro-chemical changes of valency are concerned with electrons in the outermost ring, and are reversible by loss or gain of electrons from without. Radioactive changes are concerned with changes in the inner nucleus, and such changes are not reversible by loss or gain of electrons from the ring. But it is the electrical charge of the nucleus which, primarily, controls the power of the ring to lose or gain electrons from without, and controls, therefore, its chemical and most of its physical properties.

A. van der Broek's view as to the magnitude of the intra-atomic charge has recently received a most remarkable confirmation by an entirely new method.[37] The recently discovered power of crystals to reflect X-rays has had already remarkable results in crystallography. It has also enabled the wave-length of the X-rays themselves to be accurately measured. Moseley used as anti-cathodes for the production of the X-rays a series of metals occupying the successive places of the Periodic Table between calcium and zinc, excepting scandium. In each case X-rays of two definite wave-lengths and different intensity, the characteristic type of X-rays for these elements were found to be emitted. The wave-lengths of the stronger rays alone have been dealt with as yet.

It was found that the frequency of this ray is connected by a simple relation, deduced from Bohr's theory, with a fundamental quantity which increases in units as we pass from one element to the next, and which can only be the charge of the central positive nucleus. The number so far found is one less than the atomic number in each case, 19 for calcium and 29 for zinc, instead of 20 or 30, the number of the place these elements respectively occupy in the Periodic Table. For iron, cobalt, and nickel the numbers are 27, 28, and 29, in agreement with their chemical properties and order in the Periodic Table, though, of course, the atomic weight of cobalt is greater than that of nickel, just as that of argon is greater than that of potassium, or that of tellurium greater than that of iodine. We thus have another definite proof that it is the nuclear charge, rather than the atomic mass, which fixes the position of the element in the Periodic Table.

The present ideas of the structure of atoms thus centre round two distinct regions, and it is probable that a third distinct intermediate region will have to be postulated. There is the exceedingly small nucleus, consisting of the "sub-atoms" of matter in association with positive electricity, —on the average approximately unit charge per two units of mass—and a few negative electrons only. The charge is thus overwhelmingly positive, and its variation, unit by unit, gives the successive places in the Periodic Table. Then there is the outer region of the atom, which alone is concerned in chemical and in almost all the common physical phenomena, and this consists of sufficient negative electrons to compensate the central positive charge, or one, two, three, and so on less than this in the case of the uni- or multi-valent ions of the elements. How many, or rather how few, of these electrons are directly concerned in these phenomena cannot be said, but probably, from the periodicity of atomic properties, only the outermost ring or shell of the structure. Lastly, it is necessary to differentiate an inner ring or shell of electrons, to account for the production of characteristic X-rays and some of the newer phenomena exhibited by matter.

NATURE OF THE ARGON GASES

LASTLY, the nature of the inert gases may be referred to. Although these occupy the zero family, and are without chemical affinities, it cannot be concluded that they consist of the unalterable nucleus without any external ring system of electrons. For they possess physical properties—a characteristic spectrum and volatility—and these, as well as the chemical properties, have been ascribed to the outer ring rather than to the nucleus. Moreover, although never yet detached by chemical agencies, these argon gases possess detachable electrons, for the α-particle carries a double positive charge, and in this respect is precisely analogous to the divalent ions of the alkaline-earth family of elements (I, p. 30). We may ask, therefore, why compounds of helium of the type $HeCl_2$ do not exist.

Some years ago, in connection with a new and very perfect method for purifying the inert gases by subjecting them at low pressure to the vapour of metallic calcium,' it was noticed that helium, so purified, conducted the discharge with difficulty at low pressure. As a matter of fact, X-rays and cathode-ray phenomena make their appearance in pure helium at a pressure of 0.5 mm.—at a pressure, that is, at which a common gas conducts the discharge with the maximum facility. It became important to decide whether absolutely pure helium would conduct the discharge at all. As a result of the investigation it was concluded that the difference between helium and other gases is one of degree only. About five to ten molecules of helium are equivalent to one of a common gas, so that if the discharge in helium is compared at *any* pressure with that in another gas not at the same pressure, but at one five to ten times less, the discharge phenomena are indistinguishable in the two cases.[34] Although this has never been completely explained, it is clear that we must regard the power of helium to conduct the discharge as a property of the element, and not as being due to impurities. In other words, helium and the other inert gases possess detachable electrons, capable of being detached by ordinary electrical agencies, as well as by the more extraordinary agencies at work in radioactive phenomena. In addition to the "divalent" helium ion, analogous to the α-particle, Sir J. J. Thomson, by his positive-ray method, finds a monovalent ion of mass 12, which he ascribes to $(He)_3$,[35] which may prove to be very significant.

The chemical inertness of the inert gases and the zero number of the family in the periodic grouping cannot therefore mean that these elements have no detachable electrons, but rather that the electrons are not detached by chemical agencies.

A suggestion was made by Armstrong, as long ago as 1895, that the molecules of the argon gases are made up of more than one atom with such intense affinity for one another that they are destitute of external combining power, and that in consequence we do not know either their atomic weight or position in the Periodic Table.[36] On the other hand, to-day the monatomicity of the argon gases is accepted for an immense variety of reasons, in addition to those originally available. Indeed, it is directly proved by radioactive evidence of the most varied description, both in connection with the volume occupied by the emanation and the helium produced in a known time from radium, for which the number of atoms disintegrating has been counted; by the density of the emanation, as shown by direct and indirect diffusion methods; by the fact that the stopping power of helium is in accord with Bragg's atomic square-root law, and is nearly the same as that of hydrogen, though its density is twice as great; and lastly, by the periodic law generalisation here dealt with, which confirms the zero position assigned to the inert gases. But Armstrong's suggestion contains the germ of an idea which serves to explain the chemical inertness of the argon gases. Instead of supposing that the atoms of the argon gases have an intense affinity for one another, it may be supposed that this intense and overwhelming affinity exists between the single argon atom and its detachable electrons. Just as chemists now explain many of the properties of nitrogen and its compounds as being due to a preference of the nitrogen atoms for one another in their combinations rather than for any other atom, so we may regard the chemical inertness of argon as conditioned by its preference for electricity rather than for matter. It will not part with its detachable electrons to chlorine, to form compounds of the type ACl_2, because its affinity for the unattached electron is greater than that for the electron attached to a chlorine atom, that is, for chlorion.

Considerations of this character make it clear that the figures attached to the various families of the Periodic Table are relative rather than absolute. They represent not so much the absolute number of detachable electrons in the outer atomic ring as the number which one atom will succeed in detaching from another, the absolute number present in the ring being unknown.

REFERENCES

References to the text of the previous volume are prefixed by I, thus—(I, p. 40).

[1] A. Fleck, Trans. Chem. Soc., 1913, **103**, 381 and 1052.
[2] A. S. Russell, Chem. News, Jan. 31, 1913, **107**, 49.
[3] G. von Hevesy, Physikal. Zeitsch., Jan. 15, 1913, **14**, 49.
[4] K. Fajans, *ibid.*, Feb. 15, 1913, **14**, 131 and 136.
[5] F. Soddy, Chem. News, Feb. 28, 1913, **107**, 97; Jahr. Radioakt. Electronik, 1913, **10**, 188.
[6] F. Soddy, Trans. Chem. Soc., 1911, **99**, 72.
[7] A. Schuster, Nature, Mar. 13, 1913, **91**, 30; F. Soddy, *ibid.*, Mar. 20, 57.
[8] Sir. W. Crookes, Trans. Chem. Soc., 1888, Presidential Address.
[9] G. von Hevesy, Phil. Mag., 1912 (vi.), **23**, 628; Physikal. Zeitsch., 1912, **15**, 672 and 715.
[10] *Ibid.*, Paper read at British Association Meeting, 1913; Chem. News, Oct. 13, 1913, 166.
[11] *Ibid*, Phil. Mag., 1913 (vi.), **25**, 390; with L. Putnoky, *ibid.*, 415; Physikal. Zeitsch., 1913, **14**, 49, 63, and 1202.
[11A] F. Paneth and G. von Hevesy, Sitzungsber. K. Akad. Wiss., Wien, 1913, **123** (ii. A), 1037; **122** (ii. A), 993.
[12] K. Fajans, Le Radium, 1913, **10**, 171.
[13] H. Geiger and J. M. Nuttall, Phil. Mag., 1912 (vi.), **23**, 439; Stefan Meyer and F. Paneth, Sitzungsber. K. Akad. Wiss., Wien, 1912, **121** (ii. A), 1403.
[14] K. Fajans and O. Göhring, Naturwissenschaften, April 4, 1913, 339; Physikal. Zeitsch., 1913, **14**, 877; O. Hahn and L. Meitner, *ibid.*, 752; A. Fleck, Phil. Mag., 1913 (vi.), **26**, 528.
[15] J. Satterly, Proc. Camb. Phil. Soc., 1912, **16**, 667.
[16] T. Barratt, Proc. Phys. Soc., London, 1912, **24**, 112.
[17] K. Fajans, Physikal. Zeitsch., 1911, **12**, 369; 1912, **13**, 699.
[18] K. Fajans and W. Makower, Phil. Mag., 1912 (vi.), **23**, 292.
[19] F. Soddy, Phil. Mag., 1909 (vi.), **18**, 739.
[20] E. Marsden and C. G. Darwin, Proc. Roy. Soc., 1912, **87A**, 17; O. Hahn and L. Meitner, Physikal. Zeitsch., 1912, **13**, 390.
[21] L. Meitner, Physikal. Zeitsch., 1912, **13**, 623; P. Beer and K. Fajans, *ibid.*, 1913, **14**, 947; E. Marsden and R. H. Wilson, Phil. Mag., 1913 (vi.), **26**, 354.
[22] E. Marsden and R. H. Wilson, Nature, Sept. 11, 1913, **92**, 29.
[23] L. Wertenstein, Le Radium, 1912, **9**, 6; B. Bainu and L. Wertenstein, Compt. rend., 1912, **155**, 475.
[24] A. B. Wood, Phil. Mag., 1913 (vi.), **26**, 586.
[25] R. Whytlaw-Gray, Nature, 1913, **91**, 659.
[26] O. Hönigschmid, Monatsh., 1912, **33**, 253.
[26A] H. G. J. Moseley, Proc. Roy. Soc, 1913, **88A**, 471.
[27] F. Soddy, Nature, Aug. 21, 1913, **91**, 634.
[28] O. Hahn and L. Meitner, Physikal. Zeitsch., 1913, **14**, 752.
[28A] G. N. Antonoff, Phil. Mag., 1913 (vi.), **26**, 1058; F. Soddy, *ibid.* 1914 (vi.), **27**, Jan.

[29] F. Exner and E. Haschek, Sitzungsber. K. Akad. Wiss., Wien, 1912, **121** (ii. A), 175; A. S. Russell and R. Rossi, Proc. Roy. Soc., 1912, **87A**, 478.
[30] F. Soddy, Royal Institution Lecture, May 15, 1912.
[31] Sir J. J. Thomson, Bakerian Lecture, Proc. Roy. Soc., 1913, **89A**, 1.
[32] E. Rutherford and F. Soddy, Phil. Mag., 1903 (vi.), **5**, 561.
[33] S. Russ and W. Makower, Phil. Mag., 1909 (vi.), **20**, 875; W. Makower and E. J. Evans, *ibid.*, 882; H. P. Walmsley, *ibid.*, 1913, **26**, 401.
[34] N. Bohr, Phil. Mag., 1913 (vi.), **26**, 1, 476 and 857.
[35] E. Rutherford, Phil. Mag., 1911 (vi.), **21**, 669.
[36] A. van der Broek, Nature, 1913, **92**, 373 and 476; F. Soddy, *ibid.*, 399 and 452; E. Rutherford, *ibid.*, 423.
[37] H. G. J. Moseley, Phil. Mag., 1913 (vi.), **26**, 1024.
[38] F. Soddy and T. *D.* Mackenzie, Proc. Roy. Soc., 1908, **80A**, 92.
[39] Sir J. J. Thomson, Phil. Mag., 1911 (vi.), **21**, 225.
[40] H. E. Armstrong, Graham Lecture, Trans. Roy. Phil. Soc., Glasgow, 1912, **43**, 77.

Made in United States
Troutdale, OR
11/03/2026

56844675R00037